Essentials in Profitable Egg Production
Bulletin No. 244

by New Jersey Agricultural Experiment Station

with an introduction by Jackson Chambers

IMPORTANT NOTE & DISCLAIMER

IMPORTANT NOTE :

As with all reprinted books of this age that are intended to perfectly reproduce the original edition, considerable pains and effort had to be undertaken to correct fading and sometimes outright damage to existing proofs of this title.

At times, this task can be quite monumental, requiring an almost total rebuilding of some pages from digital proofs of multiple copies. Despite this, imperfections still sometimes exist in the final proof and may detract slightly from the visual appearance of the text.

Some images may suffer from reduced quality due to anomalies in the original scan.

DISCLAIMER :

Due to the age of this book, some methods or practices may have been deemed unsafe or unacceptable in the interim years. In utilizing the information herein, you do so at your own risk.

We republish antiquarian books with no judgment or revisionism, solely for their historical and cultural importance, and for educational purposes.

Self Reliance Books

Get more historic titles on animal and stock breeding, gardening and old fashioned skills by visiting us at:

http://selfreliancebooks.blogspot.com/

introduction

Here at **Self-Reliance Books** we are dedicated to bringing you the best in *dusty-old-book-knowledge* – this time, an old book on Egg Production.

This special edition of ***Essentials in Profitable Egg Production*** was produced by the *New Jersey Agricultural Experiment Station*, and first published in 1912, making it well over one-hundred years old. It is also known as ***Bulletin No. 244***.

This book is written with egg production in the state of New Jersey in mind, but the practical information is general and suitable for all areas.

Features sections on *Possibilities of Egg Production in New Jersey, Necessity of Securing a Heavy Yield in Winter, The Breed Best Suited for Egg Production, Saving Eggs for Hatching, Brooding Methods, Feeding Baby Chicks, Essential Features in a Good Poultry House,* and more.

This old book is an essential read for all Poultry Breeders, and anybody considering taking the plunge and entering the business. This book is a *must*-have for all those interested in the historical aspect of the Poultry industry in general, and the industry in the New Jersey area specifically.

~ Jackson Chambers

State of Jefferson, April 2018

Kellerstrass Farm
Arthur Oscar Schilling
1907

Essentials in Profitable Egg Production.

TABLE OF CONTENTS.

PAGE.

Possibilities of egg production in New Jersey........................... 5

Necessity of securing a heavy yield in winter........................... 6

The four controlling factors... 6

The poultryman and his relation to the business....................... 7

The breed best suited for egg production.............................. 8

The value of careful mating and breeding and the factors to be considered
    in selecting breeders... 9

Saving eggs for hatching.. 10

Time to hatch... 10

Manner of hatching.. 11

Brooding methods 13

Immediate care after hatching... 14

Feeding baby chicks... 15

Constant selection essential.. 16

Summer requirements for growing stock................................. 17

Time of maturity.. 18

Essential features in a good poultry house............................ 20

Economical poultry house for the housing of laying hens............... 22

The value of sanitation... 25

Feeding the laying hen.. 26

Requirements of a successful poultry ration.......................... 27

Systems of feeding.. 29

Practical egg producing ration....................................... 30

The value of sprouted oats as a source of winter succulence........... 33

Summer succulence 34

Disposition of the eggs... 34

Summary 35

NEW JERSEY
AGRICULTURAL EXPERIMENT STATIONS
BULLETIN No. 244
MAY 9, 1912

Essentials in Profitable Egg Production

By HARRY R. LEWIS.

One of the largest, most profitable and rapidly growing branches of the poultry industry in New Jersey to-day is the production of market eggs. This phase of poultry keeping has outstripped the production of market poultry, in the forms of broilers and roasters, which has made the State famous in the past. Poultrymen endeavor at present, to produce a maximum amount of eggs for table purposes. Most of these products, where they are produced extensively, find their way to the New York wholesale markets, where white shelled, fresh gathered eggs always bring the top market price, and where the demand is never quite equal to the supply.

New Jersey is especially favored from the poultryman's standpoint, for it is located in the center of one of the most thickly populated sections of the United States, surrounded by the best and largest markets in the land; markets which, in order to supply the demand, must secure poultry products from the central western part of the country, hundreds of miles away. The cost of bringing these western products East is great, and on account of the long journey, their quality not always best. Hence in competition with our best products they must sell at a greatly reduced price and consequently reduced profit. All of these factors offer great opportunities to the New Jersey poultryman or farmer, who can successfully rear and manage a flock of poultry and secure from them a good egg yield during the time of year when there is a natural shortage and when the market price is therefore at its best.

If the greatest profit is to be realized from the eggs obtained they must be produced from December to April, the season during which the average farm flocks receiving little extra care and attention produce but a few. A glance at Plate 1 will show the relation which always exists between the factors of temperature and the selling price of eggs.

The above figures are averages taken from market quotations covering a period of twenty-five years. The corresponding difference has been more pronounced during the past winter than ever before. It will be seen that the highest price is usually received for eggs during the month of January, and the lowest during May. The same differences in prices are also found in a study of the marketing of certain classes of poultry sold for meat. It is evident, therefore, that it should be the first aim of a successful poultryman to study his markets and their requirements carefully, and, if he is to get the most out of his products, to produce them at a season when the price is certain to be at the best. It must be remembered that there can be little if any profit from egg production, if on account of improper conditions in care and methods of management, an egg yield of from 20 to 30 per cent. is not secured during the winter months.

CONTROLLING FACTORS.

There are four essential factors or cornerstones in profitable winter egg production, and it may be safely said that production is limited by the minimum efficiency of any one of them.

These factors are: The Man. The Birds. Their Environment. Their Food.

THE MAN.

The man managing any business enterprise is really responsible for the success or failure of that enterprise. Successful poultry keeping is no exception to the rule. Hence a few of the requisites of the man behind the flock are outlined here.

Poultry keeping is a business of details, and an efficient poultry-man should be capable of paying close attention to details. It is lack of this one quality more than of any other which has doubt-less caused most of the numerous failures in this work. The poultryman should be conscientious and careful in this work, and a good student and observer. He should make a close study of the habits and condition of his various flocks, and should try to make the individual as much as possible the unit of study. He should be willing to try new ideas, and to profit by the experience of others. Poultry keeping is to-day in its infancy as an art when compared with the other branches of animal husbandry, and new thoughts and improved methods of feeding, housing and breeding are constantly being brought to light. "New ideas are what make progress," and progress makes greater profits.

It is important to remember that success in a business requiring constant thought and attention to the various details of manage-ment calls for a natural aptitude or liking for the occupation. With this natural love for the work, and a good practical train-ing, the production of market eggs should bring a good living to the poultryman, and a satisfactory return on the money invested. Phenomenal profits should not be expected. A yearly profit per bird of from $0.75 to $2.50 is, on the average, a safe estimate. The actual returns will depend largely upon market conditions and the price received for the eggs. Where greater profits are realized than those outlined above, much stock and eggs are usually sold for breeding purposes at an increased selling price.

THE BIRDS.

The birds are the machines which are to convert the raw product, "food" into the finished product, "eggs." Without an efficient and well balanced machine, this work cannot be accomplished at a profit. The determination of the breed best adapted as a trans-former of raw material into the product which the market re-quires, should be the first consideration. The most desirable breed for any given condition is determined by two factors.

First. The extent of the business.

Second. The character of the markets.

In the first instance, where there is to be a large production, much greater than can be disposed of locally, the total product should generally be consigned to some reliable commission house in New York City, as that is the best egg market in the East. This market pays a premium of from three to eight cents per dozen for white shelled eggs the year round. For this reason the extensive poultry plant in New Jersey, producing market eggs should select one of the breeds which lay a pure white egg. The Single Comb White Leghorn fulfills this requirement more nearly than any other.

In the second place the poultryman producing usually in smaller quantities for a home market where there is no discrimination between either brown or white eggs, will get better results from choosing one of the general purpose breeds. The Plymouth Rocks, Wyandottes and R. I. Reds are good examples of this class. After birds of this class have outlived their profitableness as egg producers, their carcass brings considerable revenue when sold for meat purposes.

The mistake is often made of attempting to produce eggs from cross bred stock. In the long run, birds which are continually cross bred possess no characteristics which are superior to that of pure bred birds. The following facts, well established by experimentation, show conclusively the value of pure bred poultry over cross breeds for any purpose. Standard bred poultry should be used because there is:

1. More reliability in breeding.
2. A larger egg production from the egg breeds.
3. An improved quality of meat in the meat breeds.
4. A uniformity in size, shape and color of the eggs.
5. More attractiveness in appearance.
6. But a slightly greater initial cost than that for mongrels.
7. No greater cost of keeping.
8. A more efficient use of the food consumed.
9. A good demand for stock and eggs for breeding.
10. Always a greater selling value.

The above factors should be studied carefully before an attempt is made to produce eggs profitably from a flock which has been crossed promiscuously.

NOTE.—The question of determining the best general purpose breed for egg production depends almost entirely upon the breeding which is back of the particular strain of the breed selected. In other words, there is a greater difference in the egg producing qualities of strains of the same breed than there is between different breeds.

If one is to be uniformly successful in egg production, he should consistently try to build up his flock. This can be done by selecting only the best females from the entire flock, by mating them to good, vigorous males, and by using this special mating as a breeding pen from which all eggs for hatching should be saved. In this way one will get a continuous improvement that could not be assured if the entire flock were used to propagate the future layers. In other words, the time has come for the poultryman to pay more attention to individual birds, and less to the flock as a whole, especially when breeding for future producers. The practice of making special breeding pens in this way is bound to result in time, in larger, more vigorous birds, better layers, and hence more profit.

Great care should be exercised not to include in these special matings, for breeding purposes, any birds which have had any poultry disease. Some diseases such as Bacillary White Diarrhea (Bacterium Polorum.), are known to be directly inherited, others are known to give to the offspring an inherited weakness which makes them especially susceptible to that particular disease. This is true of tuberculosis, diarrhea, enteritis, etc. Still others, such as roup, catarrah, and other general disorders weaken the birds constitutionally, and make them unfit for breeding. Any bird once affected with disease should be conspicuously marked, so that it may never find its way into the breeding pens.

In selecting the females for the breeding pen in the spring, the following factors have been found by observation and experimentation, to be essential to heavy egg production, namely:

1. Breeding from heavy producers.
2. The use of mature birds, preferably hens, not pullets.
3. Line breeding rather than too close in-breeding or out-crossing.
4. Breeding from birds which were early producers as pullets.
5. Selecting birds which show a good vigorous constitution.
6. Selecting for egg type (long, rather deep bodies with abundant room for the organs of digestion and reproduction).
7. Selecting large specimens of the breed.
8. Selecting late moulters since they are usually the best winter layers.
9. Breeding from birds which have shown by records to be persistent layers.
10. Selecting birds that eat well, rise early and retire late, for such birds are usually heavy layers.

Without strong, well bred birds, a good egg yield is not certain, even with the most efficient management. Hence much time and thought, and if need be, money, should be expended in the improvement of the flock by mating and breeding.

The birds set aside for breeding should be so cared for as to provide for the production of fertile eggs in the breeding season, which will hatch into vigorous chicks. They should not be crowded into small unsanitary houses, nor should they be forced for a heavy egg yield during the winter. The three factors which especially favor the production of fertile eggs are exercise, which can be made compulsory by supplying most of the feed in deep litter on the floor of the pen; green food, which for want of beets or cabbage can be supplied in the form of sprouted oats; and meat scrap, supplied liberally (15%) during the breeding season, but not excessively prior to the breeding season.

The saving of eggs for hatching purposes should begin about the first of February. They should be collected at least twice a day, and should be kept at a temperature of about 45 to 50 degrees. They should not be subjected to great variations in temperature and should be placed so that they may be turned daily, and protected from the direct rays of the sun. The factor of temperature is especially important, for the germ (embryo) begins to develop at 70 degrees Fahrenheit. High temperatures start the development of the germ, but if heat is not supplied uniformly, it ultimately dies. As a result of this, many eggs are later tested out as infertile. Eggs for hatching should be selected with a view toward uniformity in shape, size and color, the idea being to select toward the ideal type of egg for the breed kept.

TIME TO HATCH.

The time of year for hatching the chicks that are to be matured for prolific winter layers should receive careful attention. The retarded hatching of the chicks is very often the direct cause of partial failure, even among experienced poultrymen, for time enough is not given the growing pullets to come to complete normal maturity before the extremely cold weather commences in the fall, which is usually about the last of November. The exact time for hatching will depend, under average conditions, upon two factors.

First. The breeds kept.

Second. The method of growing and the condition of the range.

The light, active Mediterranean breeds are much quicker growers, maturing on good range in from four to six months. They can be hatched, therefore, later than the heavier general purpose breeds which require about four to six weeks longer to reach maturity. The Leghorns can be safely hatched from the middle of April to the middle or last of May; while heavier breeds like the Plymouth Rocks, Wyandottes, R. I. Reds, &c., will do better if hatched from the middle of March to the last of April. The hatching period as given above may be modified to some extent if it is necessary to hatch three times to secure the required number of chicks.

Growing chicks that have, during the summer, an abundance of range provided with shade and green food will grow much more rapidly and more evenly than those crowded into small bare yards. The character of the range must, therefore, be considered in deciding on the time for proper hatching. If the chicks are hatched too early they are apt to go into a fall moult after laying a few eggs in the late summer, and are not likely to resume laying again until well into the winter, or after one or more of the most profitable laying months have passed. On the other hand, they should be hatched early enough to allow a sufficient time for normal maturity before the coming of cold weather in the fall.

MANNER OF HATCHING.

The manner in which the chicks are to be hatched will depend entirely on the number of young birds desired. A poultryman desiring to hatch a family flock of not more than 100 chicks can no doubt attain his end more easily and more economically by using the natural method. On the other hand, where a large number is desired, from 200 up into the thousands, artificial incubation is more practical. The latter method, if properly followed, will produce as many and as vigorous chicks as would be

NOTE.—Excessive forcing of the growing stock is often practiced for the purpose of rapidly maturing the young birds. Some birds will invariably make smaller individuals at maturity with less stamina and vitality and will be less able to resist the cold weather of the winter. Because of the stunted growth this type of bird will lay smaller eggs, which in future years will hatch smaller chicks and the general tendency of such birds will be to degenerate during the future generations, rather than to show definite improvement.

produced by the natural method. Moreover, this will be accomplished with less labor, and at any convenient time of the year. In running an incubator the directions which comes with the machine should be followed carefully, since the manufacturer is supposedly best qualified to get the most out of his particular machine. The machine or machines should be located, if possible, partially below ground, in a building which offers protection from changes in weather conditions, provides an abundance of fresh air, and assures an adequate moisture content in the air.

Success with the incubator calls for a knowledge of the principles of incubation, and of the running of the machine, including the proper care of the lamp, its daily filling, the trimming of the wick and the maintaining of the flame at a height sufficient to furnish the required amount of heat without smoking. The incubation temperature is 103 degrees Fahrenheit. It should not vary in either direction, and a rising of the temperature should be especially guarded against on or about the fourteenth day. The eggs must be turned twice a day from the third to the eighteenth, and cooled regularly, the length of time depending on the period of incubation and the temperature of the cellar. The eggs should be cooled longer as the hatch progresses; in a warm cellar in the late spring, often as long as 20 minutes during the last week of the hatch, and usually not longer than the time required for turning during the first week of the hatch. A good test is to cool the eggs down so that they still feel warm to the cheek or back of the hand.

The question of sanitation in the incubator is very important. It is desirable to wash the machine thoroughly after each hatch with a 5% solution of creolin and to allow it to air for eighteen or twenty hours. This will insure the next hatch against infection by germs of white diarrhea or of other diseases with which the former brood may have been infected, and the presence of which cannot be determined until after the chicks are removed to the brooder.

The question of supplying the proper amount of moisture to the incubator, is often as perplexing as it is important. The final results of the hatch are determined to a great extent by this factor. The amount of moisture required will vary with the season and the character of the room in which the machine is located. Experiments during the past two winters (note, see Experiment Station Report, 1911) show conclusively that lack of a sufficient amount of moisture is very detrimental to the hatch. Moisture can be supplied in two ways, *first,* by increasing the moisture content of the whole room by sprinkling the floor; and *second,* by the use of a sand tray under the eggs. The greatest amount of moisture was required in these experiments during the last week of the hatch, and it was found desirable to raise the relative humidity at this time up to an average of from 55 to 65, depending on the season. More moisture was required in the summer than in the winter.

BROODING METHODS.

In most cases the greatest loss occurs during the early brooding period, and it is a serious problem for the poultryman brooding about 1,000 chicks to decide as to the system that is to be adopted. The small outdoor brooder of fifty chick capacity does not provide maximum efficiency and economy. On the other hand, an extensive brooder house, with a central heating plant and piping becomes too expensive, unless a large number of chicks are to be brooded. It is, therefore, the problem of the poultryman to find some efficient method of brooding which will insure a better percentage of brood, and will involve less labor and less cost for equipment and fuel. If possible, a system is desired which can be used for purposes other than brooding during the non-brooding seasons.

Experiments conducted within the past few years show conclusively that the large flock method of brooding is the most efficient, and that one of the best types of brooders to use for this work is the New York State Gasoline Brooder. This type of brooder was used last year with very excellent results. There were three houses running with three broods in each house, and there were brooded over 1,200 chicks with a total loss of only 11% up to eight weeks of age. Plate 5 shows the three houses used, the general plan of the house, and also the character of construction. The burner parts can be secured complete for $10.90, and the house complete will cost for material from $25 to $32, depending on the kind of lumber used.

TABLE NO. 1.

List of Lumber for Gasoline Brooder House.

Foundation, 2—2 x 12 x 8.
Floor Joists. 4—2 x 4 x 8.
Floor (double), 65 sq. ft. ship lap, 1 x 10.
　　　　　　70 sq. ft. 3-in. flooring.
Sides and roof, 250 sq. ft. ship lap, 1 x 10.
1 x 3 dressed white pine studding and rafters.
Roofing paper, 300 sq. ft., 120 linear ft.
40 sq. ft. 1 x 10 white pine and hinges for door and hover.
2 cellar sash.

A brooder of this type can be used for many purposes. It is an efficient brooder house during the early period when it is necessary to supply artificial heat, and after this the hover can be removed. It can then be used for a developing or colony house during the summer months, at which time it can be easily moved alongside of the corn field or in the orchard, where shade will be naturally supplied. When fall arrives it can be moved, if desired, in the lee of some farm buildings and used for a laying house for small special matings of from eight to twelve hens during the winter months.

IMMEDIATE CARE AFTER HATCHING.

Improper care just at hatching time is often responsible for the lack of normal development or proper maturity of the pullets. The following practices will aid in insuring success.

The chicks should be left in the incubator for about thirty-six hours after hatching, allowing them to dry off thoroughly, and become strong enough to stand up. They should then be moved to a brooder, taking care not to chill them during the transferring, which had previously been heated to 98 degrees and on the floor of which there is an abundance of fine sand or grit and short cut alfalfa or clover.

NOTE.—The exact temperature of the brooder must be determined by the size and character of the hover, the size of the brooder compartment, and size of the brood. With the small 50-chick hovers, starting at 98 degrees and gradually lowering about 1½ to 2 degrees a week will give the best results. With the gasoline brooders or large unit brooders, the temperature should be kept high enough to provide during the first four weeks a high degree of warmth under the center of the hover. After the first few days, owing to the large brooder compartment, the chicks will be able to adjust themselves to the temperature which best meets their requirements. Under these conditions, it is the general practice to run the temperature of the gasoline brooders higher than would be required or desirable in a brooder with smaller capacity or smaller brooder compartments.

In feeding the baby chicks the following principles apply, especially where it is proposed to mature them for layers or breeders rather than for meat purposes in the form of broilers or roasters:

Principles of Baby Chick Feeding.

1. Practice retarded or limited early feeding.
2. First feed should be easily seen and nutritious.
3. An abundance of grit and shell are invaluable.
4. Fresh water is always necessary.
5. Dry cracked grains are best for the first few days.
6. Bran is important from the ash standpoint.
7. An abundance of available ash is required.
8. Feed often and sparingly for the first two weeks.
9. Avoid sloppy wet feeds.
10. Some animal protein is necessary for best growth.
11. Keep chicks busy and hungry.
12. Succulent feed in some form is essential.
13. Feed cheaper rations as the chicks get older.
14. Feed early and late each day.
15. Compel baby chicks to take abundant exercise.
16. Practice absolute cleanliness in feeding.
17. Feed to keep the chicks growing constantly.
18. Avoid pampering and unduly fussing with baby chicks.
19. Constant thought and judgment are necessary in early feeding.

The leading object in baby chick feeding is to so care for them during the first four weeks of their growth as to get them safely through the most critical period of their life with a normally developed body and a strong frame or skeleton. After this they may be safely forced for rapid meat growth if desired by the feeding of moist mashes.

A GOOD FEEDING 'PRACTICE.

Dry Cracked Grains are the Safest.

First 18 hours in brooder, supply an abundance of grit, shell and fresh water with no solid feed. On the following day, first feeding, rolled oats about 1 oz. to 25 birds. For the remainder of second day and for the next five days feed five times daily the following cracked grain ration on the floor of the brooder:

20 lbs. fine cracked corn.
25 lbs. fine cracked wheat.
5 lbs. pin-head oat-meal.
10 lbs. granulated milk. (Fine).
3 lbs. fine charcoal.

As supplemental to this ration, feed:

> Hard boiled eggs once daily.
> Sprouted oat tops twice daily. (Small amounts.)

On the seventh day start feeding wheat bran in hoppers, leaving it before them for about two hours and omit the noon feeding of grains.   From the eighth to the fourteenth days:

Wheat bran constantly in hoppers and cracked grains four times daily.

After the fourteenth day keep the following dry mash always before them, and feed grains three times daily.

> 10 lbs. wheat bran.
> 5 lbs. corn meal.
> 5 lbs. sifted ground oats.

Meat scrap 5%, to be gradually increased during the next two weeks to 8%.

The above outline is especially valuable for feeding the first eight weeks by which time the chicks will be large enough to eat whole grains and a cheaper ration should then be used.

It is very desirable during the brooding period to allow the chicks to get outdoors on the ground as often as possible, provided the ground is free from snow or water, as they will do much better and will be much hardier and more vigorous.   As soon as they can get along without artificial heat, without crowding, the same should be removed.

SELECTION.

From the time chicks are hatched until maturity, the flock should be watched with the purpose of removing any birds which show signs of weakness or lack of inherited vitality.   It has been proved by experiments that chicks which are naturally weak at

NOTE.—This will vary with the season of year and the breed kept. Usually four weeks in artificially heated brooder houses and about seven weeks in outdoor brooders when the weather has become fairly settled in spring or after April 1st, is sufficient. The artificial heat should never be turned off in outdoor brooders during the cold or variable weather. Leghorns have a tendency to huddle and crowd when it is removed, so that it is necessary to continue longer with them and to use more care when stopping it. Chicks once taught the habit of crowding, are hard to handle, and conditions should be kept so that they will never need to huddle in order to keep warm.

birth never make profitable birds to raise to maturity, either for egg production or for meat purposes. Therefore, it is a good practice to examine the young chicks when they are about a week old and to separate those which show lack of vigor. They should be kept by themselves, and developed for rapid flesh growth and be disposed of at the squab broiler age. Another sorting out of the flock is made as soon as sex can be accurately determined, or about the tenth week. All of the males should be separated and those which are not to be kept for breeders should be fed especially for meat growth. If Leghorns are kept they should be sold as soon as the market will take them for light broilers; for after that time, even if they are strong and full of vitality, every pound of meat they take on, will cost as much or more than may be realized for it in the fall.

SUMMER REQUIREMENTS FOR GROWING STOCK.

After making these two careful selections the poultryman will have a flock of strong, vigorous pullets, ten to twelve weeks old, and the problem from this time on is to mature them to good laying condition at the right time in the fall.

There are four requirements which are essential for the most economical and certain development. They are:

1. An abundance of range for the growing stock.
2. Natural shade if possible, otherwise artificial shade should be supplied.
3. Natural green food in abundance.
4. The feeding of dry mash constantly in large self-feeding hoppers.

It is impossible from the standpoint of economy and efficient growth to mature a lot of pullets in close confinement, especially on small bare yards without any natural shade and without access to an abundance of green grass. With small flocks it may be done by the expenditure of a large amount of time and by costly methods of feeding; but on a large commercial plant or under farm conditions, it is not advisable to attempt it.

The construction of a large hopper capable of holding a large quantity of feed is a great labor saver. By allowing the birds access to its contents, a better and a more satisfactory growth is obtained, and an opportunity is given them to balance the grain rations fed. This hopper should be large enough to hold several bushels of feed, sufficient for one or two weeks.

Aside from meeting the four requirements mentioned above, the poultryman should provide his birds with clean, roomy, well ventilated summer colony houses where they will not be unduly crowded, and where they will have fresh air to breathe at all times. Stunted pullets are but too often produced by lack of proper sleeping and roosting quarters during the summer growing period. The feeder should always endeavor to keep the pullets growing constantly without any check, thus doing away with any danger of retarding the time of maturity or of reducing the ultimate size and vigor of the mature birds.

TIME OF MATURITY.

Where possible it is a good practice to place the pullets in their laying quarters at least one month before the flock is expected to gain maximum maturity, for two reasons:

First. Birds are especially susceptible to changes of environment. By giving them time to get acquainted with their future home, retardation in growth is avoided.

Second. If for any reason a maturity is delayed, it is quite possible by having the birds closely under observation, to hasten or retard their ultimate maturity by the feeding of forcing or retarding mashes. In this way it is possible to bring late hatched birds to maturity from three to five weeks earlier.

It is best to have the pullets mature before cold weather begins in the fall. This will be in October for North Jersey and November for South Jersey. This means that the flock average should be established by December first, if a profitable yield is to be obtained during the rest of the winter. In other words, it will not be possible to greatly increase the egg yield after that time. An eight per cent. yield on or about December 1st will mean a low production during the winter, but a thirty per cent. yield at that time, can be kept with little trouble, well up or above that point, until the following spring, especially will this be true if proper housing and feeding conditions are furnished.

THEIR ENVIRONMENT.

If the best performance is to be expected from layers during the winter, they must be given the most congenial surroundings possible. This means, first of all, that they must be kept in a house which is suitably located and which furnishes the desired conditions at a minimum cost. The location of the plant itself should receive careful consideration, either in the case of prospective poultrymen just starting, or of old plants in which changes are being made.

Given any desirable location, the following conditions should be present. The plant should be:

1. Within easy reach of large and, if possible, a variety of markets.
2. Rapid means of communication should be available, such as rural telephone and rural free delivery.
3. The presence or absence of efficient means of transportation should be noted; adequate express and freight service is essential and the rural trolley express is very advantageous.
4. Good roads and slight grades are another point in favor of location possessing them.
5. The location should enjoy rather mild climate, free from extreme variations in temperature, frequent fogs, or prolonged periods of damp weather.
6. The soil should be porous, sandy or gravely and well drained.

The laying houses should be placed on sloping land, preferably on the southern slope with the house facing south or southeast. They will then receive the direct rays of the sun for the greatest part of the day.

Low spots should be avoided as they are apt to be damp on account of improper air drainage. Such places are usually springy and wet, thus increasing the work of cleaning the yards and runs besides increasing the danger of disease in the flock.

The buildings themselves should be situated with a view toward saving time and labor in caring for the birds. When the plant is complete it should possess a neat and attractive appearance.

The exact location of the houses must be determined by local conditions, which will naturally vary on different farms. For one thing they should be placed where they will be protected from prevailing winds and storms and where the birds can have an abundance of range and natural shade.

The exact type of house to use will depend upon the size of the flock and the system of poultry keeping. Undoubtedly the most efficient production can be realized from large flocks of from 200 to 500 kept in large units, and given extended range during the summer and kept closely confined during the winter. In the large unit system, labor is greatly reduced and the cost of housing per bird is much less than with the isolated houses of from ten to fifty bird capacity, which on the colony system are scattered over a large area. In small flocks the careful attention which each bird receives will make possible a slightly higher individual production, but the cost of this slight increase is greater than the value of the additional eggs.

It is important that the would-be poultryman know something of the requirements of the birds, and that he build accordingly. A large investment in buildings which are elaborate and costly places an exceedingly heavy burden on the poultryman. Furthermore, such elaborate buildings rarely furnish as desirable conditions as do houses constructed more cheaply, but with a direct purpose in view.

THE ESSENTIAL FEATURES IN A GOOD HOUSE.

Economy in Construction. It is not always necessary to employ only new lumber for poultry houses. Old farm buildings can often be utilized to good advantage, especially when the frame and boards are in good condition. In many cases there are buildings about the farm which with a little extra expense for material could be remodeled into efficient houses by putting down a desirable floor, cutting openings for muslin curtains, and constructing suitable sheltered roosting places.

A heavy sill and good frame are important for insuring permanence and rigidity. One of the most economical types of construction involves the boarding up of the roof and side walls with tongued and grooved materials (yellow pine seconds being satisfactory), and covering this with some good grade of roofing paper.

Convenience in Caring for the Birds. In planning the house provision should be made for as many labor saving devices as possible, such as double swinging doors between pens with friction stops, large self feeding hoppers for the dry mash which require filling but once a week, drinking vessels which are easily and quickly cleaned, nests which are easy of access, and an inside finish which can be quickly and thoroughly cleaned when necessary.

Direct Sunlight Should Reach Every Part of the House as Much of the Day as Possible. Sunlight is the best germ destroyer known, cleansing the parts of the house where it shines. It also adds warmth and makes the environment more congenial, thus acting as a tonic to the birds during the short winter days and inducing a heavier production.

Freedom From Moisture Is Essential.—The two kinds of moisture which have to be avoided in poultry houses, where the layers are to be kept in a healthy condition, consist of condensation moisture and surface soil water. The first is caused by the condensing of atmospheric moisture on the ceiling and rafters. This is usually due to lack of sufficient head room and more often to insufficient ventilation and fresh air. This condition can be corrected by substituting muslin for the glass fronts and thus insuring plenty of circulation. The second is usually seepage water, working its way under the foundation and up through the dirt and dampening the litter. This should be guarded against by proper drainage under the foundations when the house it built, and by the construction of a suitable concrete floor which, if properly made, is impervious to water.

The House Should Be Well Ventilated Without Causing Drafts to Blow Directly on the Birds. An abundant supply of oxygen is essential if the birds are to perform their normal body functions. It is especially needed where a large number are continuously crowded together in close quarters during the entire winter, as is true of most laying houses. It can best be supplied by the use of a liberal amount of muslin in the front of the house. Such curtains allow at all times fresh air to pass in and the impure air to pass out, and this change takes place without drafts or rapid movement of the air. The muslin acts as a sieve or buffer.

The Birds Should Be Given Plenty of Room for Exercise. Exercise is essential for the health of the individuals and to prevent them from taking on too much surplus fat, a condition which would be detrimental to heavy egg production. The exercise can best be provided by feeding all grain rations in deep litter on the floor. The number of birds which can be safely kept in a house of given dimensions will depend somewhat upon the breed, and largely on the experience of the poultryman caring for them.

Under average conditions it is safest for the amateur, for the one with little experience, not to crowd the birds too closely, keeping about one bird to every four and one-half or five feet of floor space. The expert, however, who thoroughly understands the needs and methods of sanitation can successfully keep as many as one bird to every two and one-half to three square feet of floor space.

Protect the Birds From Cold Without Keeping Them Too Warm. Birds will stand intense cold much better than a warm atmosphere. If the house is drafty as well as damp, the birds become subject to colds which rapidly develop into forms of roup that quickly put the birds out of laying condition. The poultry-man should so arrange the house that the temperature of the birds' bodies will be conserved when necessary during very cold weather. This can be done by the use of muslin drop curtains in front of the perches. At no time should the temperature in the house be allowed to become low enough to freeze the combs.

The House Should Be Made as Nearly as Possible Rat and Vermin Proof. Rats are often a source of great loss, caused directly by the death of young pullets, and the cost of a good concrete floor will often be saved in one year by making the house absolutely rat proof. In this way a great saving is accomplished in the feed bill, for a family of full grown rats will eat about as much dry mash as a flock of 25 laying hens. The internal construction of the house should be as plain as possible, and should offer few hiding places for lice and mites. All internal fixtures should be made movable so that they may be taken out of the house occasionally and thoroughly cleaned.

The principles outlined above can be most economically worked out to suit New Jersey conditions in a house conforming very closely to the following type. Of the six different types of roof which are used for poultry houses, the shed roof is the best, as it covers a given floor space efficiently and at a smaller cost than any other type. The following plan of a shed roof house 20 x 40 feet is especially suited to New Jersey poultry farms. Where it is desirable to keep larger units than a forty foot house will accommodate it is recommended that the length be doubled, making it 20 x 80 feet with three cross partitions (one every 20 feet), instead of only one as in the forty foot house.

The following description of the above plan shows the important features:

Specifications for the double unit house shown in Plate No. 8. The outside dimensions are 40 x 20 feet, sills to be 4 x 6, and to be bolted to a concrete foundation wall eight inches wide and twenty inches deep, which is laid on tamped cinder or crushed stone, the entire depth of the foundation trench being three feet.

The shed roof type of construction is used with nine foot studding in front and four and one-half foot studding in back. All studding and rafters are 2 x 4 hemlock or yellow pine. A 2 x 6 girder runs the length of the building supporting the rafters and itself being supported every ten feet by 4 x 4 posts, resting on concrete piers. The plates should be made of 2 x 4 material doubled and joints broken.

All outside walls and roof to be single boarded, preferably of eight or six inch tongued and grooved yellow pine; white pine can be used, but is much more expensive. The roof and back wall should be covered with a good roofing paper; all joints should be carefully lapped and cemented.

The muslin curtains in the front wall are hinged at the top and can be lifted up. The 3 x 5 glass sash are hinged at the side and open as indicated on the floor plan. One window in each pen should be so constructed that part of the wall will open when desired, thus making a combination door and window. This will greatly facilitate cleaning and filling hoppers, &c., in an extremely long house.

The dropping boards, perches, and nests are best arranged on the back wall, the perches being hinged to the wall so that they may be hooked up when cleaning, the nests being darkened by a hinged door in front which may be let down when it is desired to remove the eggs.

The dividing partition between the units is built of boards and extends from the back wall to within six feet of the front wall; the remaining space is left entirely open. This protects the birds from any drafts when on the roosts. When desired, portable light wire partitions may be used to separate the units. A large dry mash hopper should be built into this middle partition. If four or more units are built, it is only necessary to have a hopper in the center of each two units, the other dividing partition being utilized for nesting space. This hopper should be constructed on the general style as shown in Plate No. 7, with a wooden cover hinging at the center. There is an elevated platform under the muslin front which provides room for the water fountain and grit and shell hoppers.

When the house is completed concrete floor should be laid, and should consist of three distinct layers. First, a layer of about six to ten inches of cinders or coarse gravel tamped thoroughly to serve for drainage purposes to keep the soil moisture away from the bottom of the floor. Next, a rough coat of concrete about four inches thick, and over this a finished coat of two parts of sand to one of cement, troweled smooth and rounded at the corners. Where there is danger of much moisture coming up from below it is advisable to put a layer of tarred building paper between the rough and finish coat of cement. It should be nailed down with flat headed nails, and the heads of the latter should be left sticking out about one-quarter of an inch to hold the top coat.

Such a floor is moisture-proof, rat-proof, vermin-proof, and easily and quickly cleaned.

The following is a list of materials which will be required for building a double unit as shown in the working drawings, Plate No. 8:

LIST OF MATERIALS REQUIRED AND APPROXIMATE COST.

LUMBER.

Sills—6 pieces 4 x 6 by 20 feet hemlock.
Plates—8 pieces 2 x 4 by 20 feet hemlock.
Posts—2 pieces 4 x 4 by 14 feet hemlock.
2 pieces 4 x 4 by 18 feet hemlock.
Studding—9 pieces 2 x 4 by 18 feet hemlock.
4 pieces 2 x 4 by 14 feet hemlock.
Frame for nests and dropping boards—
5 pieces 2 x 3 by 16 feet hemlock.
Eight-inch tongued and grooved yellow pine boards for roof, dropping boards, walls and nests. 2200 sq. ft.
1 x 2 white pine for curtain frames and trim. 200 linear feet.
1 x 4 white pine for nests. 100 linear feet.
One bundle plaster lath for broody coop.
Nails, 10 lbs. 20 penny wire.
50 lbs. 10 penny wire.
20 lbs. 8 penny wire.

Approximate cost of the above............................ $75 54
Roofing paper, 1060 sq. ft., or 11 rolls, at $3.00.................... 33 00
Four special sash, 3 x 5 feet, at $2.00............................ 8 00
Muslin, 8 sq. yards, at 20 cents per yard......................... 1 60
Hardware, as hinges, locks, tacks, hooks and wire................. 4 75
   Foundation and floor—
Cement, 35 bags, at 50 cents...............................$17 50
Cinders or gravel, 30 yards at $1.00....................... 30 00
Sand, 5 yards... 7 50
                                   ——— 55 00
Total cost, not including labor if concrete floor is put in the house ———
   and cinders and sand have to be purchased.................... $177 89
  This gives a cost per square foot of floor space of $0.222.
  A cost per running foot of house of $4.44.
  A cost per bird, allowing 4 sq. ft. per bird, of $0.888.
  Adding labor to this at one-fourth the cost of material, the total cost is $222.36, or $1.11 per bird.

SANITATION.

Regardless of the type or construction of the laying house, if the birds are to be kept free from disease and in a vigorous condition, it is necessary to practice careful and thorough sanitation. This work naturally groups itself along three lines, namely:

The droppings should be removed from the dropping boards whenever they are wet or give off objectionable odors. When by the use of absorbents, the moisture can be kept from them they are not harmful. It will usually be found most economical to do this cleaning at least twice a week during the winter.

The litter on the floor of the house should be removed and replaced with fresh, clean material whenever it becomes wet, whenever it becomes finely ground and loses its property to hide the grain, whenever it becomes soiled and mixed with a large quantity of droppings. An inch of coarse sand should be kept on the floor, and this covered with straw or shavings to a depth of about four to six inches.

Poultry diseases, especially those of a contagious nature, can be largely prevented by spraying the interior of the house four or five times a year with a complete disinfecting solution. The following is recommended as being very effective:

     5 quarts cream of lime.
     1 pint of creolin.
     1 quart of kerosene.

This mixture should be agitated well and diluted with equal parts of water, and applied with a force pump through a spray nozzle. A thorough application of this solution will accomplish three things much more quickly and easily than if the solution were applied with a brush.

First. A good coat of whitewash will be applied, thin, well spread and put on with force, into all the cracks and crevices.

Second. The creolin will kill any disease germs which may be present in the house.

Third. The kerosene will help to kill and drive out all red mites and to a certain extent body lice. The former can be entirely controlled by this formula, and the latter by the use of a good lice powder in connection with the above solution.

Lice and mites are a severe drain on a flock of laying hens if present in any number. Care should be taken not to let them get established.

A clean house will mean more congenial surroundings and healthier birds.

THEIR FOOD.

An abundance of the food best suited to produce the greatest vigor in the laying hens, and to stimulate the reproductive organs is necessary if a profitable egg yield is to be secured. There are two objects in feeding the laying hen, namely, to provide for the maintenance of her body and, in addition, to supply the nutrients required for the manufacture of the eggs. The following figures show the composition of the bird's body and the whole egg:

	Water.	Ash.	Protein.	Fat.
Comp. of Laying hen	55.8%	3.8%	21.6%	17%
Comp. of fresh egg	65.7%	12.2%	11.4%	8.9%

Note in the above figures the relatively high proportion of water which shows the need of keeping fresh water before the birds all the time, as well as of supplying a large amount in the form of succulent feed. The high percentage of protein in the body and in the product, accounts for the necessity of feeding a relatively large amount of concentrated food to supply the nitrogenous material present. It also explains why poultrymen in feeding an exclusive corn diet cannot expect to get a profitable egg yield in the winter.

From the above table it will be seen that the food to be fed is determined in great measure by the character of the product desired. It must be remembered in feeding that the maintenance requirements of the individual must be supplied before she may be expected to yield any product.

Food Nutrients Required for Egg Production. For compounding a feeding ration it is desirable to have a knowledge of the requirements of a hen in full laying condition. Numerous feeding standards have been proposed, all of which require judgment and care in practice, to give best results. Any given ration must occasionally be modified to meet existing conditions.

The following standard derived by W. P. Wheeler, of the N. Y. State Experiment Station, gives these requirements in detailed form:

REQUIREMENTS OF RATIONS FOR HENS IN FULL LAYING.

Digestible Nutrients Per Day for Each 100 Pounds Live Weight.

	Total Dry Matter lbs.	Ash lbs.	Protein lbs.	Carbohydrates lbs.	Fat lbs.	Fuel Value Cal.	Nutritive Ratio.
Hens of 5 to 8 lbs. weight,	3.30	.20	.65	2.25	.20	6 240	1:4.2
Hens of 3 to 5 lbs. weight,	5.50	.30	1.00	3.75	.35	10,300	1:4.6

Requirements of a Successful Ration. The compounding and balancing of poultry rations requires a complete understanding of the essential factors in poultry feeding, and these can be briefly outlined as follows:

The Ration Must Contain Sufficient Food Nutrients. The bird which is under-fed, and has no surplus fat on her body cannot be expected to do any more than provide for the actual maintenance of her body. It is, therefore, essential that the laying hen should receive an abundance of feeding material, if anything, a little in excess of the exact amount required; this excess will be stored up in her body for future use.

The Food Nutrients Must Be in the Right Proportion. The food materials, apart from being supplied in abundance, must be supplied in the right proportion if they are to be utilized to the best advantage by the bird. This proportion is expressed by the term, nutritive ratio, and means the relation of the amount of protein to the carbohydrates plus 2¼ fat. In a laying ration, this ratio should be relatively narrow, about 1 part of protein to every

4.4 parts of carbohydrates and fat, if the best results are to be secured.

The Ration Must Be Succulent and Palatable. A succulent feed is one which contains an abundance of vegetable juices, as for instance, mangel beets, cabbage, sprouted oats, &c. Experiments carried on last year, show that the feeding of succulent material during the winter, materially increases egg production. The rations fed should be palatable, thereby inducing the birds to eat more, thus encouraging assimilation, and a heavier production. The palatability of a food not only increases the amount which is consumed, but also increases the proportion which the birds can assimilate. It thus makes the ration more economical.

The Ration Must Have Sufficient Bulk. Some fiber is necessary in the ration to properly distribute the food eaten, thus distending the digestive organs and enabling the digestive juices to act more efficiently.

Heavy, compact or soggy concentrated foods should be avoided, for they have a tendency to upset the digestive system, which will immediately impair the efficiency of the bird for production.

The Feeds Must Be Economical and Not Cheap. The cheapest food materials are not always economical. The accurate way of purchasing feeds consists in finding out the amount of digestible food material present, and determining the actual cost of a pound of such material. Usually this cost is based on a pound of digestible protein, because that nutrient is the most expensive to purchase; the one which is required always in sufficient quantity, and the one which cannot be produced as abundantly on the farm.

The Ration Must Be Regularly and Intelligently Fed. Regularity in the care of poultry is of extreme importance as they are especially susceptible to any change in the routine of management, and any irregularity will be reflected in a reduced egg yield. Therefore, a definite time should be decided upon for the feeding of the various rations, and the general routine outlined should be followed very closely. Successful feeding requires the constant exercise of intelligent thought and judgment, for there arise constantly new conditions, and from the standpoint of economy and efficiency, the rations must be varied slightly to meet these changed conditions. No one ration is the best under all conditions or at all times. The prices of the food materials are constantly changing; likewise weather conditions and the age of the birds, all of which require corresponding changes in the ration.

Special Feeds Like Grit, Shell, Charcoal and Salt Must Be Supplied. Grit should be always before the birds in self-feeding hoppers. This material enables them to grind the grain and to make it more easily acted upon by the digestive juices. It also supplies some ash for formation of bone. Crushed oyster shell should be kept before them constantly, as it is one of the most efficient sources of mineral matter, aiding in the formation of the egg shell and providing ash for the frame work of the body.

Charcoal acts as a cleanser or purifier, and it is well to supply it in the dry mash to the extent of from 2 to 5%.

Salt in the ration increases the palatability and also increases assimilation by aiding diffusion, and where the birds seem to be off their appetite or for some reason it is desired to make them eat a great amount of dry mash, salt may be supplied to the extent of about 5 oz. to 100 lbs. feed.

The factors mentioned above should be borne in mind, and care should be used in compounding rations to have these factors as nearly ideal as possible, if the greatest efficiency is to be obtained from the food given.

SYSTEMS OF FEEDING.

In general the system of feeding best suited for egg production will be a dry mash, which from the standpoint of economy, should be supplied in large self-feeding hoppers. As supplemental to this dry mash, cracked and whole grains should be supplied at least twice a day in deep litter on the floor of the house.

The wet mash system of feeding is not as efficient, for it requires more labor, and produces, if not properly fed, diarrhea and digestive disorders. Its use has been almost entirely abandoned in favor of the former method.

PRACTICAL EGG PRODUCING RATION.

The following system of feeding for egg production is a result of extended experimental work at the New Jersey State Experiment Station. In the use of this or of any other ration, it must be distinctly understood that there is no such thing as a best ration, for with different breeds, under various conditions and environments, there will need to be a variety of feeding. Hence, any ration which is giving good success should be continued until some other one can be tried out in an experimental way.

The following is the New Jersey State Dry Mash, and the supplemental rations which are designed for the complete feeding of laying hens throughout the winter, together with what modifications are necessary for summer feeding.

TABLE NO. 2.

Mixture No. 1.

Dry Mash.

KIND OF FOOD.	Amount by Weight. Lbs.	Amount by Measure. Qts.	Dry Matter.	Ash or Mineral Matter.	Protein.	Carbohydrates Plus Fat x 2¼.	Cost.
Wheat Bran	200	380	176.0	11.6	24.2	90.6	$3 20
Wheat Middlings	200	240	176.0	7.6	25.6	121.4	3 50
Ground Oats	200	200	178.0	6.0	18.4	113.6	3 30
Corn Meal	100	95	89.0	1.5	7.9	76 4	1 65
Gluten Meal	100	80	92.0	.8	25.8	65.6	1 70
*Meat Scrap (H. G.).......	100	86	89.3	4.1	66.2	31.1	3 00
Short Cut Alfalfa...........	100	200	92.0	7.4	11.0	42.3	1 60
Total	1,000	1,381	892.3	39.0	179.1	541.0	$17 95
Average to one pound,	......	1.38	.892	.039	.179	.541	$.018

Nutritive Ratio, 1—3.02.

Keep this mash before the birds all the time in large self-feeding hoppers. The hoppers used should be large enough so that one filling will last from one to two weeks at the least.

During the moulting season or the months of July, August and September, it is advisable to substitute oil meal for the gluten in the same proportion, to hasten the growth of feathers. As soon as the birds get out on green grass, the alfalfa can be gradually omitted; also meat scraps are gradually reduced in amount as soon as the birds get out on free range, and can find insects and grubs. The extent to which the above mash can be cut during the summer will depend upon the character and amount of range which the birds have during that time.

NOTE.—The quality of the different brands of meat scrap is very variable, and should a lower grade with less protein and more fat be used it would raise the nutritive ratio slightly.

The above dry mash is designed especially for the feeding of White Leghorns. Where heavier breeds are kept such as Plymouth Rocks or Wyandottes, especially yearling or two year old hens, the tendency will be to take on an excess of fat. Under these conditions it is the best policy to restrict the amount of mash eaten by leaving the hopper open during the afternoon only, thus inducing the birds to work more for the cracked grains fed in the litter.

The following modification of the above mash will be found very economical for summer feeding, the change from one to the other being made gradually as soon as the birds are on free range with plenty of natural forage.

TABLE NO. 3.
Mixture No. 1A.

Summer Dry Mash.

KIND OF FOOD.	Amount by Weight. Lbs.	Amount by Measure. Qts.	Dry Matter.	Ash or Mineral Matter.	Protein.	Carbo-hydrates Plus Fat x 2¼.	Cost.
Wheat Bran	200	380	176.0	11.6	24.2	90.6	$3 20
Wheat Middlings	100	120	88.0	3.8	12.8	60.7	1 75
Ground Oats	100	100	89.0	3.0	9.2	56.8	1 65
Gluten Meal	50	40	46.0	.4	12.9	32.8	85
Meat Scrap	25	21	22.3	1.0	16.5	5.0	75
Total	475	561	421.3	19.8	75.6	243.9	$8 20
Average to one pound,	......	1.18	.887	.04	.158	.513	$ .017

Nutritive Ratio, 1—3.22.

As supplemental to the dry mash, the following scratching ration of whole grain is fed every morning, both winter and summer, about 9 o'clock, in deep litter. Its primary object, aside from its nutritive value, is to induce exercise. About 5 pounds of the scratching ration is fed to each 100 birds on the floor of the house or under some shelter, where the litter is dry and where there is protection from cold winds.

TABLE NO. 4.

Mixture No. 2.

Scratching Ration.

KIND OF FOOD.	Amount by Weight. Lbs.	Amount by Measure. Qts.	Dry Matter.	Ash or Mineral Matter.	Protein.	Carbo-hydrates Plus Fat x 2¼.	Cost.
Wheat	100	53	90	1.8	10.2	73.0	$2 20
Clipped Oats	100	98	89	3.0	9.2	56.8	1 93
Total	200	151	179	4.8	19.4	129.8	$4 13
Average in one pound,	......	.755	.889	.024	.097	.049	$.0206

Nutritive Ratio, 1—6.6.

At 4 to 5 o'clock in the afternoon, depending on the season, a night ration is fed, composed of whole grains and cracked grains at the rate of 10 pounds to each 100 birds.

TABLE NO. 5.

Mixture No. 3.

Night Ration.

KIND OF FOOD.	Amount by Weight. Lbs.	Amount by Measure. Qts.	Dry Matter.	Ash or Mineral Matter.	Protein.	Carbo-hydrates Plus Fat x 2¼.	Cost.
Cracked Corn	200	120	178	3.0	15.8	152.8	$3 30
Wheat	100	53	90	1.8	10.2	73.0	2 20
Clipped Oats	100	98	89	3.0	9.2	56.8	1 93
Buckwheat	100	66	87	2.0	7.7	53.3	2 00
Total	500	337	444	9.8	42.9	335.9	$9 43
Average in one pound,	......	.674	.888	.019	.085	.071	$.018

Nutritive Ratio, 1—7.8.

It will be noted that this manner of feeding gives to the birds the materials suitable for supplying the heat to the body during the night. The above night ration is designed for White Leghorns; when feeding heavier breeds, it is desirable to eliminate one-half of the cracked corn and to substitute barley for the buckwheat. During the summer months, a night ration of equal parts of cracked corn, wheat, oats and barley will supply the requirements; the amounts to be fed, depending on the amount and condition of the range.

A good rule in feeding the night ration is to feed all that the birds will eat, or rather more, so as to have a little left for them to go to work on in the morning. A good feeder will occasionally go among the birds at night when they are on the perches and will feel their crops. If they are not full early in the evening he will conclude that either the layers are not getting enough or that they have lost their appetite. In either case, the defect should be immediately corrected.

One special advantage of the dry mash system outlined above is the fact that each bird is allowed to balance her own ration according to her particular requirements and tastes.

Twenty birds of average weight, if fed the above ration, will receive during the winter months the following food nutrients per day:

Lbs.	Protein.	C. H. plus fat.	Cost.
8.0	1.1	4.87	$0.14

The feeding of some succulent material in addition to this ration cannot be too strongly recommended.

The following method of sprouting oats was found to be the most successful. The oats should be soaked in water at a temperature of from 60 to 70 degrees F., for about forty-eight hours in pails or galvanized wash tubs, and during this soaking process there should be added from five to ten drops of formalin to kill the spores of moulds and to insure a clean, sweet feed. After soaking they are spread out about one inch thick on trays, which are placed in a sprouting rack, seven to each rack, the trays being ten inches apart, and kept at a temperature of from 60 to 80 degrees.

In from seven to ten days, depending on the temperature, they will have developed sprouts about three to four inches long, as well as a massive root development, the entire mass being very tender and succulent. The birds will eat this ravenously. About one square inch of feeding surface is supplied daily to each bird, or what they will clean up quickly. The oats cannot be fed in excess as they are laxative, and are apt to produce diarrhea. The rack shown in Plate No. 9 has a capacity for approximately 500 hens if kept working constantly, one tray for each day in the week.

SUMMER SUCCULENCE.

For the poultryman who is compelled by lack of space or other causes to closely confine his birds during the summer, it will be found very profitable to divide the run or yard given them into two, and to rotate the crops, allowing the birds first to feed in one yard, and then in the other. By planting seasonable crops, like peas and oats, peas and barley, buckwheat, millet, cow peas, and late in the summer, such crops as vetch, crimson clover and wheat or rye, the birds will have a continuous supply of green food during the summer and green crop to feed on early in the spring. If these crops are allowed to make about four inches to six inches of growth before the birds are turned in on them, they will not become woody nor will the birds clean them up in a day or so, but they will supply the best of green food for about four weeks, or while another crop is growing. This method not only supplies the succulent feed in the cheapest and best form, but also purifies the runs and keeps them in a clean, healthy condition, which is an important item when a lot of birds are given restricted range.

DISPOSITION OF EGGS.

The poultryman who can successfully market his eggs after they are produced will realize much more than the one who has not that knack or who has not made a study of market conditions. In order to get the most for the product, it is essential that the producer study market conditions and requirements, and if possible, cater to or meet them in as many ways as possible. Nearly all markets, retail and wholesale alike, will pay a premium for a guaranteed strictly fresh article, also for products which are graded according to size and color and are uniform in shape. Most markets will pay an increased price for eggs put up in a neat and attractive manner, in a substantial thirty dozen case with ordinary fillers. In some instances it is found profitable to use the one dozen cartons, and by grading and guaranteeing the eggs at home, where they are produced, the poultryman with an extensive output can soon build up a demand for his particular brand which will insure him a good market at relatively high prices as compared with general market quotations.

SUMMARY.

The production of market eggs as a business is attaining constantly greater proportions in New Jersey each year. It is the object of this bulletin to describe in a popular and instructive form, factors which should receive consideration in this type of poultry keeping, and to assist in a greater production at the season of the year when prices are at their best. Briefly stated, the factors of special importance are:

The selection of a well bred, producing strain, of the breed best suited to meet market requirements, which on plants of extensive production should be Single Comb White Leghorns.

The practicing of careful mating and breeding each year to insure healthy, vigorous offspring for future layers.

The hatching of the pullets early enough to give them time for normal maturity before extreme cold weather in the fall. (April and May.)

The constant selection for vigor from birth to maturity and the elimination and disposal of all weak birds.

The keeping of the growing stock healthy and inducing a continuous growth, by giving free range, shade and green food in abundance.

The hatching of the chicks under proper conditions and brooding them in large flocks.

The placing of the pullets which are to be kept for winter layers in their permanent quarters in October.

The providing of a suitable house which should be of the open front curtain type; providing light, fresh, air, room for exercise, protection from cold and winds, freedom from dampness, and at the same time a house which is economical in construction and convenient for the attendant in caring for the birds.

The practicing of thorough and frequent sanitation, especially during the winter when the birds must be closely confined, by spraying the interior of the house occasionally with a complete disinfecting solution.

The providing of the birds with a sufficient amount of the right kind of food material. The best system of feeding being the hopper feeding of dry mash, supplemented by cracked and whole grain in deep litter.

The providing of a continuous supply of succulent feed (mangel beets, cabbage or sprouted oats).

The studying of market conditions and requirements and attempting by catering to them to get the best market for the eggs produced.

The making of a continuous and careful study of the business as a business, and attempting to work out the most efficient and economical method of management possible.

Attention to details of marketing pays.

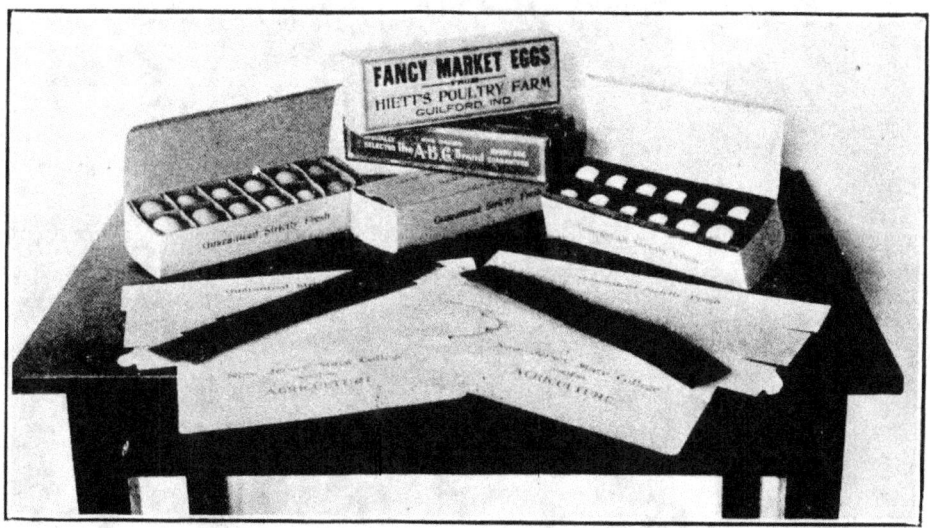

Desirable types of cartons and substantial well stenciled shipping case.

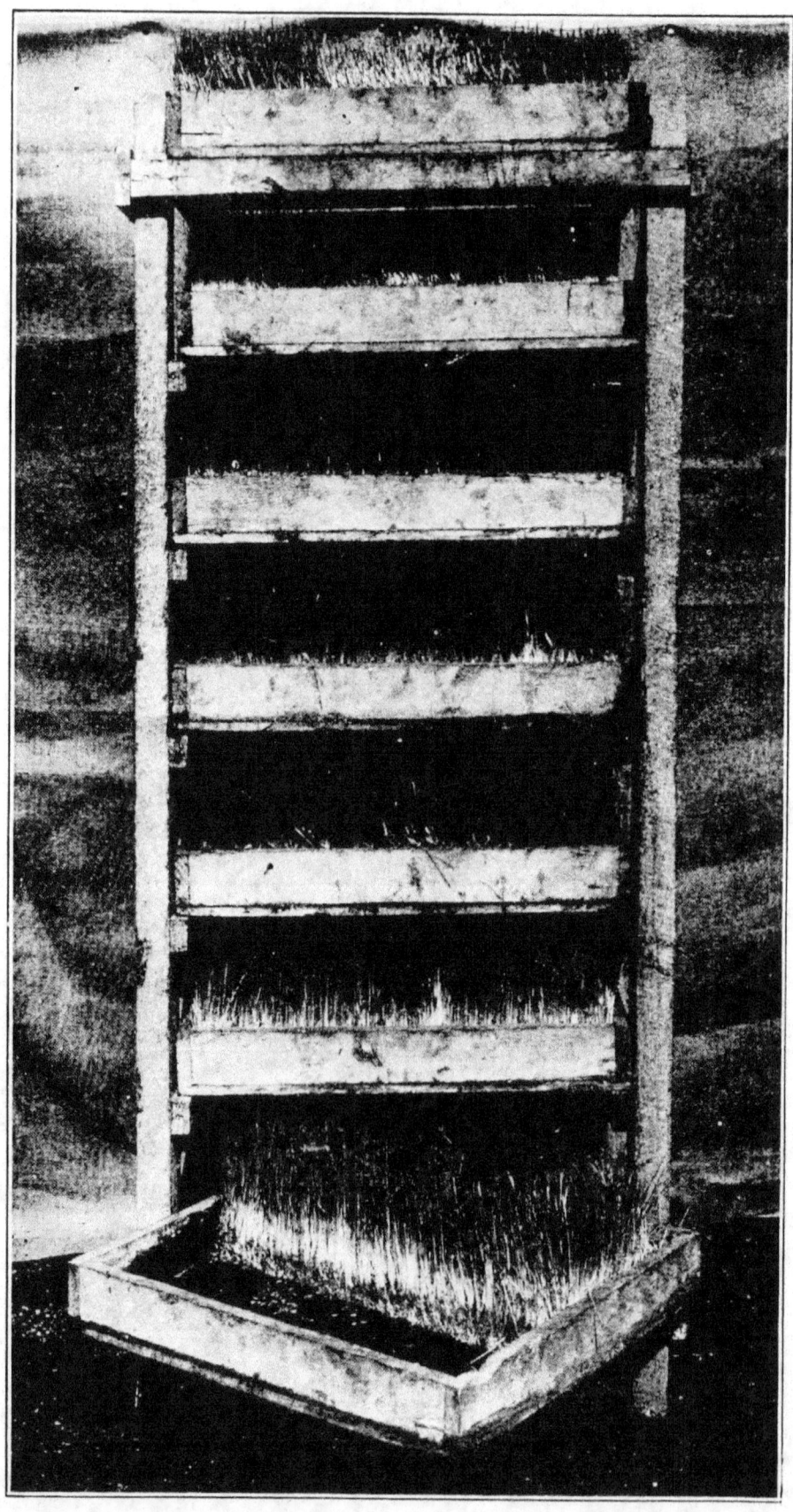

Sprouting rack and condition of oats at feeding time in bottom tray.

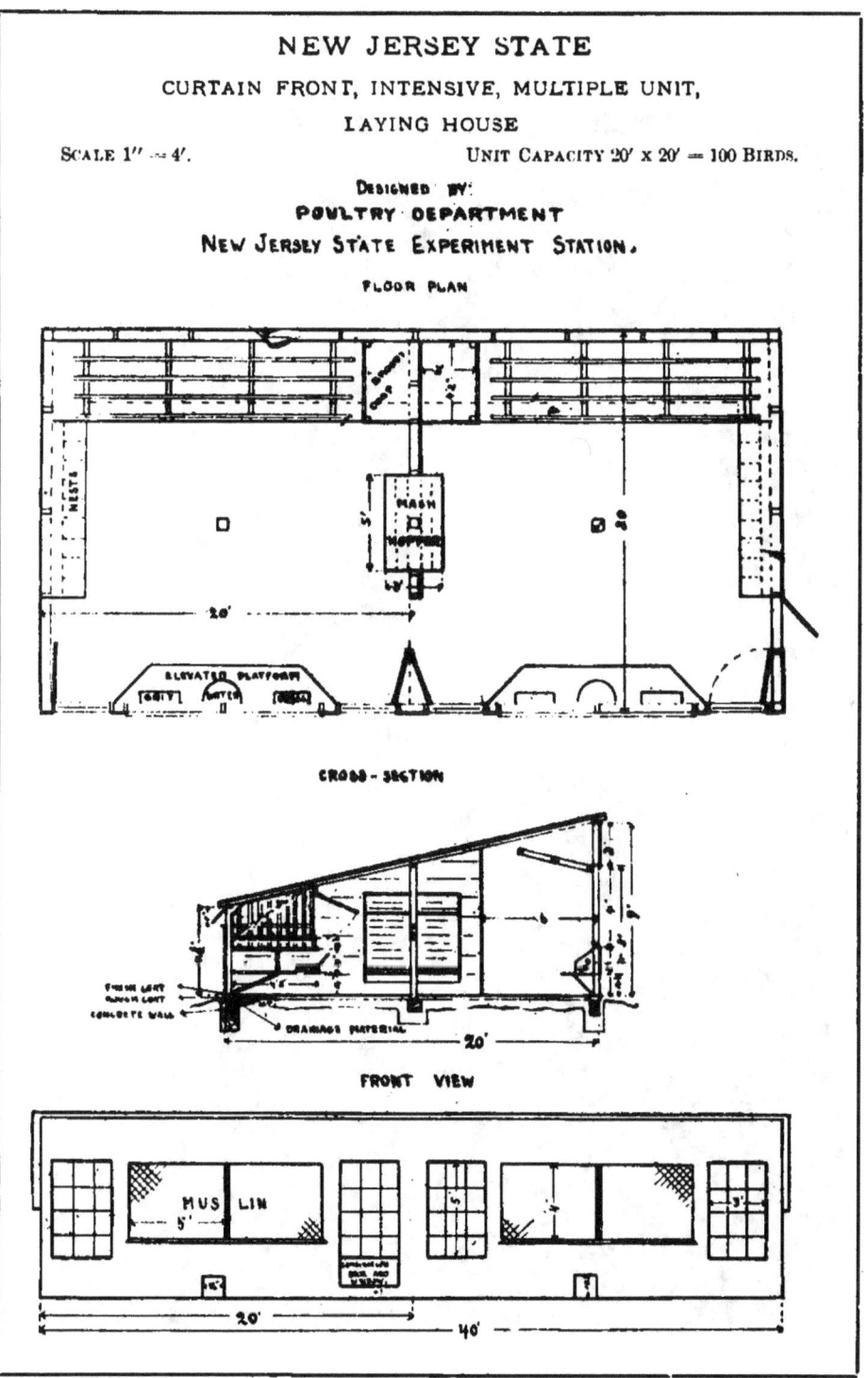

NEW JERSEY STATE

CURTAIN FRONT, INTENSIVE, MULTIPLE UNIT,

LAYING HOUSE

SCALE 1″ = 4′.　　　　　　　　UNIT CAPACITY 20′ x 20′ = 100 BIRDS.

DESIGNED BY:
POULTRY DEPARTMENT
NEW JERSEY STATE EXPERIMENT STATION.

FLOOR PLAN

CROSS-SECTION

FRONT VIEW

An economical poultry house for farm poultry keeping or intensive poultry plants.

PLATE No. 7.

New Jersey State Colony Dry Mash Hopper.

An inexpensive hopper capable of greatly reducing the cost of labor in feeding the growing stock on the range.

PLATE No. 6.

Lot A.

Lot B.

These two lots of pullets were selected at one week of age. Lot (A) showing signs of inherited strength and vigor, while lot (B) showed at the same time lack of these qualities. Note the respective condition of both flocks in November when they were coming to maturity.

PLATE No. 5.

New York State Gasoline Brooders.

Note light but strong construction, insuring easy and safe moving to and from range.

PLATE No. 4.
Constitutional Vigor in Breeding Males.

Strong.

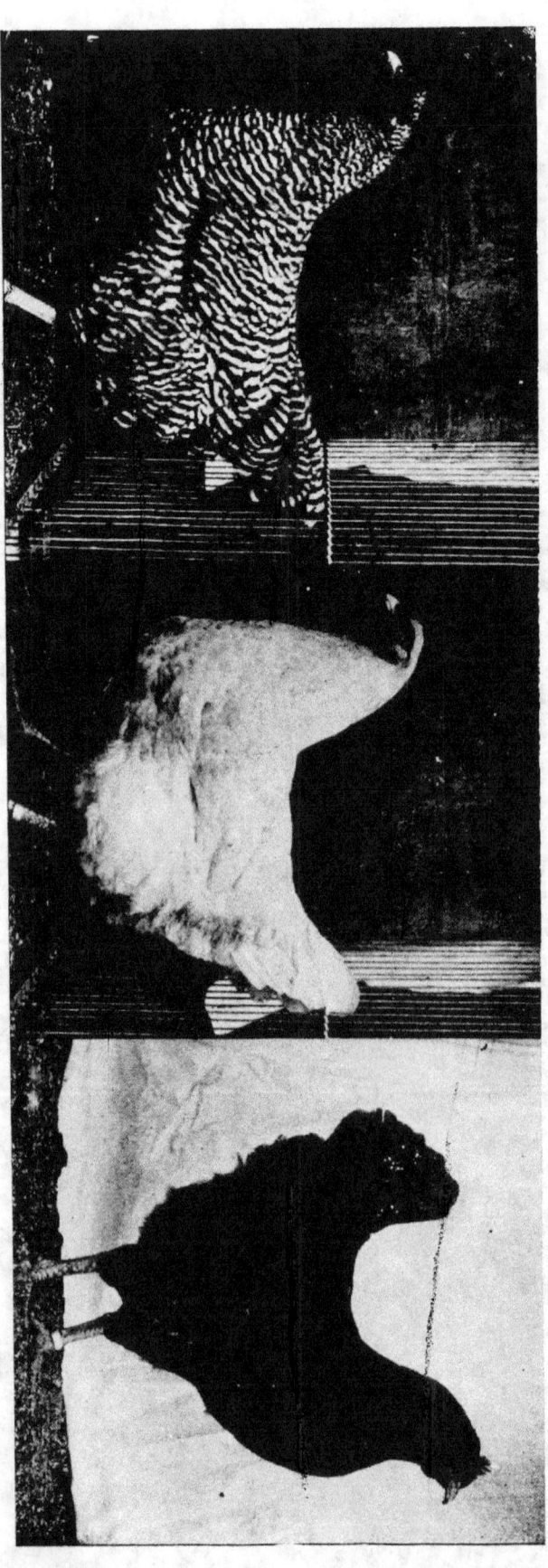

PLATE No. 3.

Barred Plymouth Rock.

White Wyandotte.

S. C. Rhode Island Red.

Three excellent general purpose breeds, showing utility types.

A utility type of Single Comb White Leghorn, just coming into maturity. As the laying period progresses the tail will become contracted and will be carried lower. Note long body and active appearance.

TEMPERATURE & PRICE CURVES

MARKET EGGS
WHOLESALE

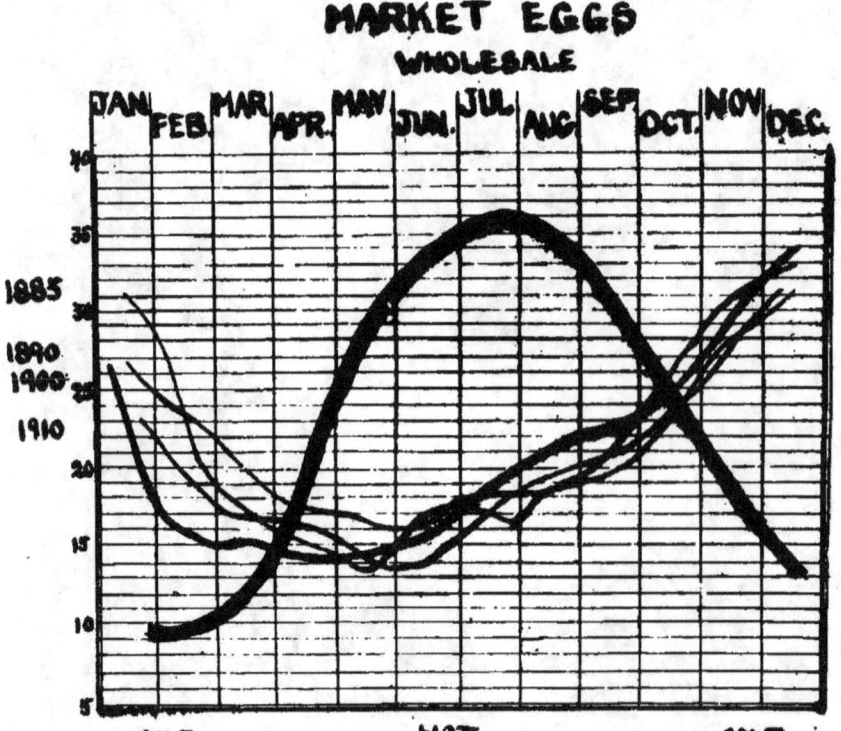

The heavy black line shows the average yearly variation in temperature, while the smaller lines show the average prices paid for fresh eggs in the wholesale market throughout the year.